This book dedicated to the ....

This book dedicated to my respected sir Dr. Pranab kumar Dan

PREFACE

This book addresses optimization theory and analysis in the context of an overall effort to achieve quality. It is designed for use as a primary book as a supplementary text in a material/mechanical/Industrial engineering course, and as a resource for academic and practical approach.

The main characteristics of this book are:
•It assumes that the reader's goal is to achieve a suitable balance of cost, schedule, and quality. It is not oriented toward critical systems.

•It presents modern optimization techniques suitable for near-term application, with sufficient technical background to understand their domain of applicability and to consider variations to suit technical and organizational constraints.

CONTENTS

Optimiazation approches in industrial enginedering

1. Optimization and Decision Theory
2. Introduction Of MCDM Methods
3. Selection Approaches In MCDM Methods With Example
4. IE in mechanical design optimization.
5. Concept of fuzzy logic-An example
6. Other optimization technique

OPTIMIAZATION APPROCHES IN INDUSTRIAL ENGINEDERING

1.1 Introduction of Optimization and Decision Theory

It is hard to imagine any science or engineering discipline, where optimization and/or decision theoretical models are not of major concern. Telecommunication systems, engineering design, logistics networks, manufacturing plants, biological and financial systems -to name a fewrely heavily on optimization approaches. For instance, an important design problem in structural mechanics is solved by finding the optimal solution of the corresponding equilibrium problem. Moreover, companies can reduce their costs significantly by applying optimization techniques in order to design and operate their supply chains.

The Optimization and Decision Theory research are mainly involves the development of algorithms based on linear, nonlinear, integer, dynamic and stochastic programming techniques. Frequently, optimal algorithms require excessive computational resources and time when applied to major problems encountered in real life. In such cases, it is crucial to develop effective heuristic algorithms by exploiting the problem structure. Hence, research in this Group employs both classical heuristics and metaheuristics, which have become increasingly popular recently.

The set of problems studied in the group includes, but is not restricted to: machine tool selection, energy efficient routing in wireless sensor and ad hoc networks, global optimization, constrained optimization and project scheduling.

If in machine designing, the material and geometrical parameters are optimized simultaneously then it is common to assume empirical formulas approximating a relation between material parameters for example the bending fatigue limit (Sbf) and ultimate tensile strength (UTS) as a function of hardness. If the choice of material is limited to a list of pre-defined candidates, then two difficulties can be appeared. First, a discrete optimization process should be followed against material parameters. Second, properties of different alternatives materials may not indicate any obvious correlation in the given list. The main goal is to choose material with best characteristic among alternatives.

1.2 Introduction Of MCDM Methods

Multiple criteria decision making (MCDM) is the process of selecting the best alternative from a set of feasible alternatives considering multiple conflicting criteria. In precise terms criteria are considered to be 'strictly' conflicting if the increase in satisfaction of one results in a decrease in satisfaction of the other. An MCDM process always contains at least two alternatives and two conflicting criteria (Bhattacharya et al., 2003). MCDM are divided two broad categories: Multiple

Attribute Decision Making (MADM) and Multiple Objective Decision Making (MODM). Several useful tools for solving of MCDM problems are

- Simple Additive Weighting method (SAW)
- Technique for Order Preference by Similarity to Ideal Solution (TOPSIS)
- Multi Objective Optimization Ratio Analysis(MOORA)
- Analytical Hierarchy Method (AHP)
- Analytical Network Method ANP etc.
  - Here discussed some example tool of MCDM theory
  - **SIMPLE ADDITIVE WEIGHTING (SAW)**

Step 1 Formation of decision matrix: Criterion outcomes of decision alternatives can be collected in a table called Decision Matrix comprised of a set of columns and rows. The matrix rows represent decision alternatives, with matrix columns representing criteria. A value found at the intersection of row and column in the matrix represents a criterion outcome - a measured or predicted performance of a decision alternative on a criterion. The decision matrix is a central structure of the MCDA/MCDM since it contains the data for comparison of decision alternatives.

$$X = \begin{matrix} & C_1 & & C_J & & C_n \\ A_1 \\ \vdots \\ A_i \\ \vdots \\ A_m \end{matrix} \begin{bmatrix} x_{11} & \cdots & x_{1j} & \cdots & x_{1n} \\ \vdots & \cdots & \vdots & \cdots & \vdots \\ x_{i1} & \cdots & x_{ij} & \cdots & x_{in} \\ \vdots & \cdots & \vdots & \cdots & \vdots \\ x_{m1} & \cdots & x_{mj} & \cdots & x_{mn} \end{bmatrix}$$

(1)

x_{ij} is the performance rating of alternative i with respect to criterion j,

A_j is i^{th} alternative, C_j is the jth criterion

Step 2 Formation of Weight Matrix

Different importance weights to various criteria may be awarded by the decision makers. These importance weights forms the weight as follows.

$$W = [W_1 \cdots W_j \cdots W_n]$$
(2)

Step 3 Normalization of performance rating

Units and dimensions of performance ratings of columns under criteria differ. For the purpose of comparison, these performance ratings are converted into dimensionless units by normalization using following equations

$$\bar{x}_{ij} = \frac{x_{ij}}{\max_i(x_{ij})} \text{ for benefit criteria } j$$
(3)

$$\bar{x}_{ij} = \frac{\min_i(x_{ij})}{x_{ij}} \text{ for non-benefit criteria } j$$
(4)

Normalized decision matrix

$$\overline{X} = \begin{matrix} A_1 \\ A_2 \\ \vdots \\ A_m \end{matrix} \begin{bmatrix} \overline{x}_{11} \cdots & \cdots \overline{x}_{1j} \cdots & \overline{x}_{1n} \\ \overline{x}_{i1} \cdots & \cdots \overline{x}_{ij} \cdots & \overline{x}_{in} \\ \vdots & \vdots & \vdots \\ \overline{x}_{m1} & \overline{x}_{mj} & \overline{x}_{mn} \end{bmatrix}_{m \times n}$$

(5)

Step 4 composite score: Computation of composite score (CS_i) for alternative i

$$CS_i = \sum_{j=1}^{n} \left(\overline{w}_j * \overline{x}_{ij} \right)$$

Step 5 Ranking and selection of best alternative: Ranking of products in descending order of composite scores (CS_i).

> ➤ **Technique for Order Preference by Similarity to Ideal Solution (TOPSIS)**

TOPSIS is an evaluation method that is often used to solve MCDM problems [2, 3].

It has a number of applications [4, 5] in practice, such as comparison of company performances, financial ratio performance within a specific industry and financial investment in advanced manufacturing systems, etc. However, there are also some limits to it. So far, the work on how to improve original TOPSIS method has mainly emphasized on improving the weight to sensitize the R value [6, 7]. Besides, there has also been improvement on formula of the R value, such as the 'Miqiezhi' method [8]. Because of the complexity of evaluation problems, a better and simpler method is required to understand the inherent relationship between the R value and alternative evaluation. In this report, a novel, modified TOPSIS (M-TOPSIS) method is described as a process of calculating the distance between the alternatives and the reference points in the D+ D−-plane and constructing the R value to evaluate quality of alternative.

♣ Algorithm of TOPSIS method under MCDM

The idea of TOPSIS can be expressed in a series of steps:

Step1 All the original criteria receive tendency treatment. We usually transform the cost criteria into benefit criteria, which is shown in detail as follows ;

(i) The reciprocal ratio method (X ij = 1/X ij), refers to the absolute criteria;

(ii) The difference method (X ij = 1 −X ij), refers to the relative criteria.

After tendency treatment, construct a matrix

$$X' = [X'_{ij}]_{n \times m}, i = 1, 2 \ldots, n; \quad j = 1, 2 \ldots, m. \tag{2.1}$$

Step2 Calculate the normalized decision matrix A. The normalized value aij is calculated as

$$A = [a_{ij}]_{n \times m}, \quad a_{ij} = X'_{ij} / \sqrt{\sum_{i=1}^{n} (X'_{ij})^2} \quad i = 1, 2 \ldots, n; \quad j = 1, 2 \ldots, m. \tag{2.2}$$

Step3 Determine the positive ideal and negative ideal solution from the matrix A.

$$A^+ = (a_{i1}^+, a_{i2}^+, \ldots, a_{im}^+), a_{ij}^+ = \max_{1 \leqslant i \leqslant n}(a_{ij}), \quad j = 1, 2 \ldots, m \tag{2.3}$$

$$A^- = (a_{i1}^-, a_{i2}^-, \ldots, a_{im}^-), a_{ij}^- = \min_{1 \leqslant i \leqslant n}(a_{ij}), \quad j = 1, 2 \ldots, m \tag{2.4}$$

Step4 Calculate the separation measures, using the n-dimensional Euclidean distance. The separation of each alternative from the positive ideal solution is given as:

$$D_i^+ = \sqrt{\sum_{j=1}^{m} W_j(a_{ij}^+ - a_{ij})^2} \qquad (2.5)$$

Similarly, the separation from the negative ideal solution is given as

$$D_i^- = \sqrt{\sum_{j=1}^{m} W_j(a_{ij}^- - a_{ij})^2} \qquad (2.6)$$

Step5 For each alternative, calculate the ratio Ri as:

$$R_i = \frac{D_i^-}{D_i^- + D_i^+} \quad i = 1, 2 \ldots, n \qquad (2.7)$$

Step 6 Rank alternatives in increasing order according to the ratio value of Ri in step5.

> ➤ MULTI OBJECTIVE OPTIMIZATION RATIO ANALYSIS (MOORA)

The MOORA method which was introduced by Brauers (Brauers, 2006) is such a multi objective optimization technique that can be successfully applied to solve various types of MCDM problems.

- Algorithm of MOORA method under MCDM

The MOORA method starts with a matrix of responses (performance measures) of different alternatives on different criteria (objectives or attributes). The matrix is shown below (Equation 1).

$$X = \begin{matrix} & C_1 & \cdots & C_j & \cdots & C_n \end{matrix} \\ \begin{matrix} A_1 \\ \vdots \\ A_i \\ \vdots \\ A_m \end{matrix} \begin{bmatrix} x_{11} & \cdots & x_{1j} & \cdots & x_{1n} \\ \vdots & \cdots & \vdots & \cdots & \vdots \\ x_{i1} & \cdots & x_{ij} & \cdots & x_{in} \\ \vdots & \cdots & \vdots & \cdots & \vdots \\ x_{m1} & \cdots & x_{mj} & \cdots & x_{mn} \end{bmatrix}$$

(6)

Where x_{ij} is the performance rating (response) to the ith alternative (A_i) under jth criterion (C_j). m is the number of alternatives and n is the number of criteria.

The MOORA method employs a ratio system in which each response of an alternative on an attribute (criterion) is compared to a denominator. The denominator is a representative for all alternatives concerning that attribute (Brauers et al. 2007; Kalibatas and Turskis, 2008).

Brauers et al. (2008) considered various ratios such as the square root of the sum of squares of each alternative per objective, total ratios, Scharlig ratios, Weitendorf ratios, Jutter ratios, Stop ratios, Van Delft and Nijkamp ratios of maximum value, Korth ratios, Peldschus et al. and Peldschus ratios for nonlinear normalization. They concluded that the square root of the sum of squares of each alternative per objective is the best one for the denominator which is given below.

$$x_{ij}^* = \frac{x_{ij}}{\sqrt{\sum_{i=1}^{m}(x_{ij}^2)}}$$

(7)

x_{ij}^* is normalized value of response *i* with respect to attribute *j*. In the current research work, the maximum score under each attribute has also been used as the denominator of the ratio system and an effort has been made to exhibit that this ratio system is also suitable for finding the optimal solution. The following ratio system is the second best for normalization process in MOORA.

$$x_{ij}^* = \frac{x_{ij}}{\max_i (x_{ij})}$$

(8)

For the computation of normalized response using the above Eq. (2b), first the maximum score under each attribute is found. Then all the scores under certain attribute irrespective of benefit or non-benefit are divided by the concerned maximum score using Eq. (2b). x_{ij}^* is a dimensionless quantity in the interval [0,1] representing the normalized score of alternative *i* on attribute *j*. However, sometimes the interval could be [-1; 1]. For example in the case of productivity growth of some factories, industries, sectors, regions or countries may be negative instead of positive thus the interval becomes [-1;1] (Brauers *et al.*, 2008).

For multi-objective optimization these normalized performances are added in case of maximization and subtracted in case of minimization. Then the optimization problem becomes

$$y_i^* = \sum_{j=1}^{g} x_{ij}^* - \sum_{j=g+1}^{n} x_{ij}^*$$

(9)

Where *g* is the number of benefit criteria to be maximized and *(n-g)* is the number of non-benefit criteria to be minimized. y_i^* is final score of *i*th alternative with respect to all the attributes. In the above case it is assumed that all the attributes are of same importance.

$$y_i^* = \sum_{j=1}^{g} w_j * x_{ij}^* - \sum_{j=g+1}^{n} w_j * x_{ij}^*$$

(10)

Where w_j^* is the weight of jth attribute (criterion), which can be evaluated using any well-known approach either AHP or Entropy method. The value of y_i^* may be positive, negative or zero. These y_i^* values are arranged in descending order. The best alternative is one which is associated with highest y_i^* value and the worst alternative is one which is associated with the lowest y_i^* value.

➤ ENTROPY

Entropy was originally a thermodynamic concept, first introduced into information theory by Shannon (see Shannon, 1948 [21]). It has been widely used in the engineering, socioeconomic and other fields. According to the basic principles of information theory, information is a measure of system's ordered degree, and the entropy is a measure of system's disorder degree.

Step1 Calculate p_{ij} (the ith scheme's jth indicator value's proportion).

$$p_{ij} = r_{ij} / \sum_{j=1}^{m} r_{ij}, \quad r_{ij}$$ is the ith scheme's jth indicator value.

Step2 Calculate the jth indicator's entropy value

$$e_j. \; e_j = -k \sum_{i=1}^{m} p_{ij} \ln p_{ij}, \; k = 1/\ln m, \; m$$ is the number of assessment schemes.

Step3 Calculate weight wj (jth indicator's weight).

$$w_j = (1 - e_j) / \sum_{j=1}^{n}(1 - e_j), n \text{ is the number of indicators,}$$

$$\text{and } 0 \leq w_j' \leq 1, \sum_{j=1}^{n} w_j = 1.$$

In entropy method, the smaller the indicator's entropy value ej is, the bigger the variation extent of assessment value of indicators is, the more the amount of information provided, the greater the role of the indicator in the comprehensive evaluation, the higher its weight should be.

1.3 Present Approaches In MCDM Methods With Example

Optimal design of gear or any other machine requires the consideration of the two type parameters known as material and geometrical parameters. The choice of stronger material parameters may allow the choice of better geometrical parameters and vice versa. Very important difference among these two parameters is that the geometrical parameters are often varied independently. On the other hand, material parameters can be inherently correlated to each other and may not be varied independently. An example of which being the variation of the bending fatigue limit (Sbf) with the core hardness (HB) for some steel materials. If these parameters would be varied independently in an optimization case, it may result in infeasible solutions. Therefore, the final choice of material may not be possible within available data base.

MATERIAL PERFORMANCE INDICES

The main characteristics considered in the design of gears are:
- ✓ surface fatigue limit (Ssf),
- ✓ root bending fatigue limit (Sbf),
- ✓ wear resistance of tooth's flank
- ✓ High tensile strength to prevent failure against static loads
- ✓ High endurance strength to withstand dynamic loads
- ✓ Low coefficient of friction
- ✓ Good manufacturability

Generally cast iron, steel, brass and bronze are preferred for manufacturing metallic gears with cut teeth. Where smooth action is not important, cast iron gears with cut teeth may be employed. Commercially cut gears have a pitch line velocity of about 5 metre/second. For velocities larger than this, gear sets with non-metallic pinions as one member are used to eliminate vibration and noise. Non-metallic materials are made of various materials such as treated cotton pressed and moulded at high-pressure, synthetic resins of the phenol type and rawhide. Moisture affects rawhide pinions. Gears made of phenolic resins are self-supporting on the other hand other two types are supported by metal side plates at both ends of the plate. Large wheels are made with fretting rings to save alloy steels. Wheel centre is commonly cast from cast iron. The ring is forged or roll expanded from steel of the respective grade specified by the tooth design.

Problem Definition

An organization has got 9 different materials with different specifications for gear. The decision maker considered 7 selection criteria. The materials are as follows:

SL. NO.	Material	GRADE
Material 1	Cast iron	SAE J431-43500
Material 2	Ductile iron	EN-GJS 418
Material 3	S.G. iron	BS 2789
Material 4	Cast alloy steel	BS 2795
Material 5	Through hardened carbon steel	SAE 4140
Material 6	Surface hardened alloy steel	SAE 8620
Material 7	Carburised steel	SAE 8620
Material 8	Nitrided steel	EN40B
Material 9	Through hardened carbon steel	817M40

Table-01

The selection criteria are as follows:

C1	Surface Hardness (Bhn)
C2	Core Hardness (Bhn)
C3	Surface Fatigue Limit (MPa)
C4	Bending Fatigue Limit (MPa)
C5	UTS (MPa)
C6	Cost (INR) Per kg
C7	Supply Lead Time (In week)

Table-2

Out of 7 criteria, 5 criteria viz. C1: Surface Hardness (Bhn), C2: Core Hardness (Bhn), C3: Surface Fatigue Limit (MPa),C4: Bending Fatigue Limit (MPa),C5: UTS (MPa) are beneficial criteria because their higher values are desirable and remaining viz. C6: Cost (INR) Per kg, C7: Supply Lead Time(In week) are non-beneficial criteria because their lower values are desirable.

The objective of the decision maker is to assess the performance of the materials. Counseling the above 7 criteria to ultimately select the best material. The decision maker applied SAW, TOPSIS and MOORA methods for their simplicity, adaptability, applicability and is of applications. The decision matrix for the materials with respect to the criteria shown below:

Formation of decision matrix

Table: Suggested materials and their properties in a gear material selection problem[A]

MATERIAL	Grade	Surface (Bhn) Hardness (C1)	Core (Bhn) Hardness (C2)	Surface Fatigue Limit (MPa) (C3)	Bending Fatigue Limit (MPa) (C4)	UTS (MPa) (C5)	Cost (INR) Per kg (C6)	Supply Lead Time (In week) (C7)
Cast iron (M1)	SAE J431-43500	200	200	330	100	380	55	2
Ductile iron (M2)	EN-GJS 418	220	220	460	360	880	55	2
S.G. iron (M3)	BS 2789	240	240	550	340	845	47	3
Cast alloy steel (M4)	BS 2795	270	270	630	435	845	66	4
Through hardened carbon steel (M5)	SAE 4140	270	270	670	430	620	58	5
Surface hardened alloy steel (M6)	SAE 8620	542	229	1160	680	1850	60	6
Carburised steel (M7)	SAE 8620	647	297	1500	920	2300	60	5
Nitrided steel (M8)	EN40B	693	297	1250	760	1250	72	5
Through hardened carbon steel (M9)	817M40	185	185	500	430	635	74	5

Table-3

^{A}Data(except material grade,cost and supply lead time) are taken form Hofmann (1990) where Vickers hardness values have been converted to Brinell values using conversion tables in http://www.gordonengland.co.uk/hardness/brinell_conversion_chart.htm

^{A}Data (**material grade,cost and supply lead time**) are taken form Bill Forge Private Limited (Plant I)9C, **Bommasandra Industrial Area,Hosur Road,Bangalore - 562 158,India**

MATLAB

MATLAB supports a variety of graphs that enable you to present information effectively. The type of graph you select depends, to a large extent, on the nature of your data. The following list can help you select the appropriate graph:

- ✓ Bar and area graphs are useful to view results over time, comparing results, and displaying individual contribution to a total amount.
- ✓ Pie charts show individual contribution to a total amount.
- ✓ Histograms show of data values.
- ✓ Stem and stair step plots display discrete data.
- ✓ Compass, feather, and quiver plots display direction and velocity vectors.
- ✓ Contour plots show equivalued regions in data.
- ✓ Interactive plotting enables you to select data points to plot with the pointer. Animations add an addition data dimension by sequencing plots.

Computational result by MATLAB:

ENTROPY METHOD:

RESULT:

ENTROPY METHOD							
criteria	C_1	C_2	C_3	C_4	C_5	C_6	C_7
weighted values	0.1635	0.1129	0.1634	0.1290	0.1143	0.1336	0.1833

SAW METHOD									
Material	M_1	M_2	M_3	M_4	M_5	M_6	M_7	M_8	M_9
The values of (s)	3.3105	3.9933	3.9247	3.7710	4.0601	4.9866	6.1170	5.2557	3.0018
Arranging the final value in descending order:-				M7 > M8 > M6 > M5 > M2 > M3> M4 > M1 > M9					

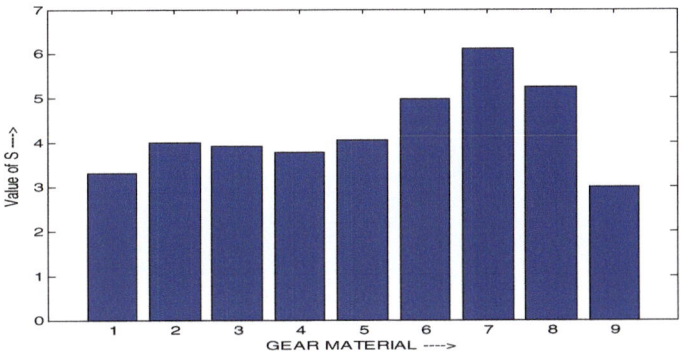

MOORA METHOD:

RESULT:

STEP 1   Determination of normalized decision matrix

	C_1	C_2	C_3	C_4	C_5	C_6	C_7
M_1	0.1623	0.2685	0.1258	0.0597	0.1000	0.2990	0.1538
M_2	0.1785	0.2953	0.1754	0.2149	0.2316	0.2990	0.1538
M_3	0.1948	0.3222	0.2097	0.2029	0.2224	0.2555	0.2308
M_4	0.2191	0.3625	0.2402	0.2596	0.2224	0.3588	0.3077
M_5	0.2191	0.3625	0.2555	0.3223	0.3131	0.3153	0.3846
M_6	0.4398	0.3074	0.4423	0.4058	0.4868	0.3262	0.4615
M_7	0.5250	0.3987	0.5720	0.5491	0.6052	0.3262	0.3846
M_8	0.5623	0.3987	0.4767	0.4536	0.3289	0.3914	0.3846
M_9	0.1501	0.2484	0.1907	0.2566	0.1671	0.4023	0.3846

STEP 2   Determination of weighted normalized decision matrix:

	C_1	C_2	C_3	C_4	C_5	C_6	C_7

M_1	0.0268	0.0301	0.0203	0.0075	0.0113	0.0409	0.0287
M_2	0.0295	0.0331	0.0283	0.0270	0.0261	0.0409	0.0287
M_3	0.0322	0.3222	0.2097	0.2029	0.2224	0.2555	0.2308
M_4	0.2191	0.3625	0.2402	0.2596	0.2224	0.3588	0.3077
M_5	0.0362	0.0406	0.0413	0.0404	0.0353	0.0431	0.0717
M_6	0.0727	0.0344	0.0715	0.0509	0.0548	0.0446	0.0860
M_7	0.0868	0.0446	0.0924	0.0689	0.0682	0.0446	0.0717
M_8	0.0929	0.0446	0.0770	0.0569	0.0371	0.0535	0.0717
M_9	0.0248	0.0278	0.0308	0.0322	0.0188	0.0550	0.0717

STEP 3: Determination of weighted multi objective optimization:

(the value of a is the sum of all weighted normalized values for all beneficial column)

Material	M_1	M_2	M_3	M_4	M_5	M_6	M_7	M_8	M_9
The values of (a)	0.0960	0.1439	0.1526	0.1732	0.1938	0.2843	0.3609	0.3085	0.1344

The value of b is sum of all weighted normalized values for all non-beneficial column

Material	M_1	M_2	M_3	M_4	M_5	M_6	M_7	M_8	M_9
The values of (b)	0.0696	0.0696	0.0779	0.1064	0.1148	0.1306	0.1163	0.1252	0.1267

Material	M_1	M_2	M_3	M_4	M_5	M_6	M_7	M_8	M_9
The values of (a-b)	0.0264	0.0744	0.0747	0.0668	0.0790	0.1537	0.2446	0.1833	0.0077

| Arranging the final value in descending order:- | M7 > M8 > M6 > M5 > M3 > M2 > M4 > M1 > M9 |

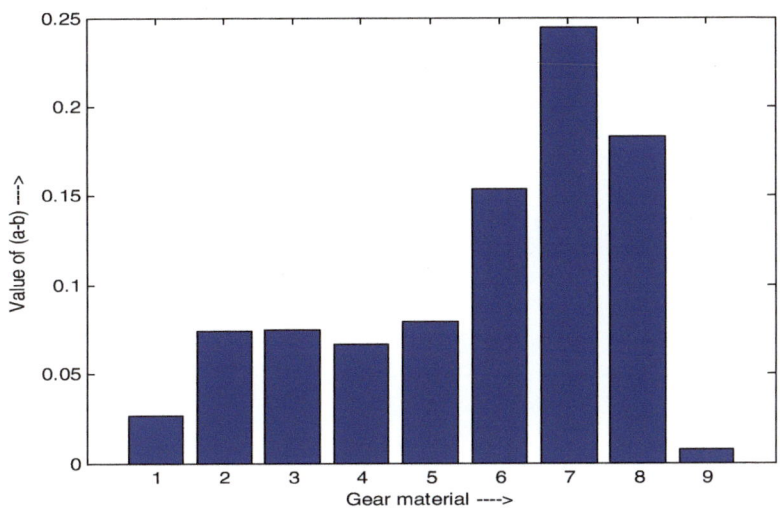

Fig:7

TOPSIS METHOD BY USING MATLAB:

Material	M_1	M_2	M_3	M_4	M_5	M_6	M_7	M_8	M_9
The values of R_i	0.3286	0.3944	0.3273	0.2967	0.3508	0.5560	0.6905	0.5941	0.1932
Arranging the final value in descending order:-				M7 > M8 > M6 > M2 > M5 > M1 > M3 > M4 > M9					

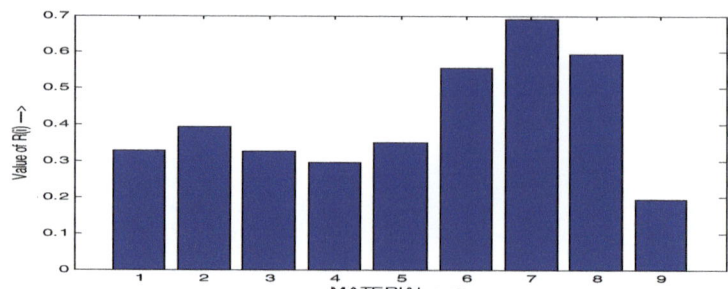

Fig:8

Comparative analysis of ranking of gear materials using MCDM methods:

MATERIAL	SAW (RANK)	MOORA (RANK)	TOPSIS (RANK)
M1	8	8	6
M2	5	6	4
M3	6	5	7
M4	7	7	8
M5	4	4	5
M6	3	3	3
M7	1	1	1
M8	2	2	2
M9	9	9	9

Table-4

DISCUSSION:

From the result we see that for the three different process of MCDM, the result is almost same. The ranking of 1st, 2ND, 3RD and 9th Materials are same for those three different processes. For the simplicity, prompt result getting the accurate value and also getting the best ranking we have used the MATLAB software. By this software we can also make rank of any system for any number of alternatives and criteria within a fraction of second with accuracy.

CONCLUSION

It is quite clear that selection of a proper Gear Materials for a given manufacturing application involves a large number of considerations. The use of SAW, TOPSIS and MOORA methods are observed to be quite capable and computationally easy to evaluate and select the proper material from a given set of alternatives. These methods use the measures of the considered criteria with their relative importance in order to arrive at the final ranking of the alternative Gear Materials. Thus, these popular MCDM methods can be successfully employed for solving any type of decision-making problems having any number of criteria and alternatives in the manufacturing domain. Use of MATLAB software makes MCDM problem simple and gives prompt results which is very essential in today's decision making environment.

1.4 IE in mechanical design optimization

Materials and process selection are key issues in optimal design of industrial products. Substituting and selecting materials for different machining parts is relatively common and often. Material selection is a difficult and subtle task, due to the immense number of different available materials. From this point of view paper deal with a set of major gear design criteria which are used for gear material selection. The paper introduces the decision models for best selective material processes. These models are explored in terms of fatigue theory as well as product life cycle is explained and their optimization problems are discussed analytically. The fatigue life of product merged on the best selection of material for better product development, considering economic aspect of present situation. In this book, the writer dives deeper into optimization technique to find out product life.

OVERVIEW OF GEAR MATERIAL:

Gears are commonly made of cast iron, steel, bronze, phenolic resins, acetal, nylon or other plastics. The selection of material depends on the type of loading and speed of operation, wear life, reliability and application. Cast iron is the least expensive. ASTM / AGMA grade 20 is widely used. Grades 30, 40, 50, 60 are progressively stronger and more expensive. CI gears have greater surface fatigue strength than bending fatigue strength. Better damping properties enable them to run quietly than steel.

Nodular cast iron gears have higher bending strength together with good surface durability. These gears are now a days used in automobile cam shafts. A good combination is often a steel pinion mated against cast iron gear. Steel finds many applications since it combines both high strength and low cost. Plain carbon and alloy steel usage is quite common.

Through hardened plain carbon steel with 0.35 - 0.6% C are used when gears need hardness more than 250 to 350 Bhn. These gears need grinding to overcome heat treatment distortion. When compactness, high impact strength and durability are needed as in automotive and mobile applications, alloy steels are used. These gears are surface or case-hardened by flame hardening, induction hardening, nitriding or case carburizing processes. Steels such as En 353, En36, En24, 17CrNiMo6 widely used for gears.

Bronzes are used when corrosion resistance, low friction and wear under high sliding velocity is needed as in worm-gear applications. AGMA recommends Tin bronzes containing small % of Ni, Pb or Zn. The hardness may range from 70 to 85Bhn. Non metallic gears made of phenolic resin, acetal, nylon and other plastics are used for light load lubrication free quiet operation at reasonable cost. Mating gear in many such applications is made with steel. In order to accommodate high thermal expansion, plastic gears must have higher backlash and undergo stringent prototype testing.

1.3 GEAR MATERIAL SELECTION MODELS:

Optimal design of gears requires the consideration of the two type parameters:Material and geometrical parameters. The choice of stronger material parameters may allow the choice of finer geometrical parameters and vice versa. Very important difference among these two parameters is that the geometrical parameters are often varied independently. On the other hand, material parameters can be inherently correlated to each other and may not be varied independently. An example of which being the variation of the bending fatigue limit (Sbf) with the core hardness (HB) for some steel materials. If these parameters would be varied independently in an optimization case, it may result in infeasible solutions. Therefore, the final choice of material may not be possible within available data base.If gear material and geometrical parameters are optimized simultaneously then it is common to assume empirical formulas approximating a relation between material parameters for example the bending fatigue limit (Sbf) and ultimate tensile strength (UTS) as a function of hardness. If the choice of material is limited to a list of pre-defined candidates, then two difficulties can be appeared. First, a discrete optimization process should be followed against material parameters. Second, properties of different alternatives materials may not indicate any obvious correlation in the given list. The main goal is to choose material with best characteristic among alternatives.
Table 1. Shows suggested nine materials with their characteristics in a gear material selection .

RESEARCH AGENDA:

In an industry, design is a field that generally deals with different practices of design parameters; the research and development of processes, machine and equipment. The materials and process selection are key issues in optimal design of industrial products. Substituting and selecting materials for different machining parts is relatively common and often. Material selection is a difficult and subtle task, due to the immense number of different available materials. From this point of view paper deal with a set of major gear design criteria which are used for gear material selection. The main gear design criteria are: surface fatigue limit index, bending fatigue limit index, Surface Hardness, Core Hardness, Ultimate tensile strength,Cost,supply Lead Time. Using computer allows a large amount of information to be treated rapidly. One the most suitable models, for ranking alternatives gear materials, are SAW, MOORA, TOPSIS which using a multiple criteria, which all material performance indices and their uncertainties are accounted for simultaneously.

Industrial engineering is a branch of engineering dealing with the optimization of complex processes or systems. It is concerned with the development, improvement, implementation and evaluation of integrated systems of people, money, knowledge, information, equipment, energy, materials, analysis and synthesis, as well as the mathematical, physical and social sciences together with the principles and methods of engineering design to specify, forecast, and evaluate the results to be obtained from such systems or processes.

This paper concerns about increment the decision of material selection of gear manufacturing process and improvement the machinability, accuracy, quality, optimize the cost and time with the industrial view. Overall improvement of optimal design of a gear in manufacturing process considering the fatigue life and other aspect of materials.

1.5 OVERVIEW OF MCDM:

Multiple criteria decision making (MCDM) is the process of selecting the best alternative from a set of feasible alternatives considering multiple conflicting criteria. In precise terms criteria are considered to be 'strictly' conflicting if the increase in satisfaction of one results in a decrease in satisfaction of the other. An MCDM process always contains at least two alternatives and two conflicting criteria (Bhattacharya et al., 2003). MCDM are divided two broad categories: Multiple Attribute Decision Making (MADM) and Multiple Objective Decision Making (MODM). Several useful tools for solving of MCDM problems are,

- Simple Additive Weighting method (SAW)
- Technique for Order Preference by Similarity to Ideal Solution (TOPSIS)
- Multi Objective Optimization Ratio Analysis(MOORA)
- Analytical Hierarchy Method (AHP)
- Analytical Network Method (ANP) etc.

1.5.2 TECHNIQUE FOR ORDER PREFERENCE BY SIMILARITY TO IDEAL SOLUTION (TOPSIS)

TOPSIS is an evaluation method that is often used to solve MCDM problems [2, 3]. It has a number of applications [4, 5] in practice, such as comparison of company performances, financial ratio performance within a specific industry and financial investment in advanced manufacturing systems, etc. However, there are also some limits to it. So far, the work on how to improve original TOPSIS method has mainly emphasized on improving the weight to sensitize the R value [6, 7]. Besides, there has also been improvement on formula of the R value, such as the 'Miqiezhi' method [8]. Because of the complexity of evaluation problems, a better and simpler method is required to understand the inherent relationship between the R value and alternative evaluation. In this report, a novel, modified TOPSIS (M-TOPSIS) method is described as a process of calculating the distance between the alternatives and the reference points in the D+ D−-plane and constructing the R value to evaluate quality of alternative.

1.5.4 ENTROPY

Entropy was originally a thermodynamic concept, first introduced into information theory by Shannon (see Shannon, 1948 [21]). It has been widely used in the engineering, socioeconomic and other fields. According to the basic principles of information theory, information is a measure of system's ordered degree, and the entropy is a measure of system's disorder degree.

1.6 0VERVIEW OF MATLAB

MATLAB is a high-performance language for technical computing. It integrates computation, visualization, and programming in an easy-to-use environment where problems and solutions are expressed in familiar mathematical notation. Typical uses include:

• Math and computation

• Algorithm development

• Modeling, simulation, and prototyping

• Data analysis, exploration, and visualization

• Scientific and engineering graphics

• Application development, including graphical user interface building

MATLAB is an interactive system whose basic data element is an array that does not require dimensioning. This allows you to solve many technical computing problems, especially those with matrix and vector formulations, in a fraction of the time it would take to write a program in a scalar non interactive language such as C or Fortran.

The name MATLAB stands for matrix laboratory. MATLAB was originally written to provide easy access to matrix software developed by the LINPACK and EISPACK projects. Today, MATLAB uses software developed by the LAPACK and ARPACK projects, which together represent the state-of-the-art in software for matrix computation.

MATLAB features a family of application-specific solutions called toolboxes. Very important to most users of MATLAB, toolboxes allow you to learn and apply specialized technology. Toolboxes are comprehensive collections of MATLAB functions (M-files) that extend the MATLAB environment to solve particular classes of problems. Areas in which toolboxes are available include signal processing, control systems, neural networks, fuzzy logic, wavelets, simulation, and many others.

OBJECTIVE OF WORK

3.1 The proposed research work is planned into 5 stages:

    3.1.1 Identification of problem and setting up objective.
    3.1.2 Analysis of parameter and design of optimization tool.
    3.1.3 Effective simulation using MATLAB.
    3.1.4 Find fatigue life by construction of S-N diagram
    3.1.5 With the help of optimization tool rank the best alternatives (materials).

4.1.1 Phase 1:
Different objectives would be chosen form literature review for analysis and improvement, such as various alternative of gear materials and their criteria. Formation of MATRIX of gear materials to improve machinability, accuracy, quality, optimize the cost and time with the industrial view of a gear for better manufacturing product.

4.1.2 Phase 2:
Design is considered with proper selection of tool. For this research work MCDM is preferred as an optimization tool. But since the various method of MCDM is heavily used in material selection problem solving, hence might be some possible effective method can be added in this paper. In this stage more objective and harder matrix will be taken together.
A model of the proposed technique is presented below as flow diagram. This technique is a multiple criteria based decision making optimization technique which is mainly based on ranking to solve the problem and indicate the best selection of gear material.

4.1.3 Phase 3:
In this paper the problem solved by the MATLAB and showing the graph of materials and also detect the accuracy of the following problem.

4.1.3 Phase 4:
Find fatigue life by construction of S-N diagram

4.1.4 Phase 5:
With the help of optimization tool (SAW, MOORA & TOPSIS) rank the best alternatives (materials), by MATLAB software.

4. Selection of Gear Materials Considering Technical Economic and Supply Aspect by Ranking in MATLAB

4.1 PROBLEM DEFINITION
An organization has got 03 different materials with different specifications for gear. The decision maker considered 7 selection criteria. The materials are as follows:

MATERIAL	Grade	Surface (Bhn) Hardness (C1)	Core (Bhn) Hardness (C2)	Surface Fatigue Limit (MPa) (C3)	Bending Fatigue Limit (MPa) (C4)	UTS (MPa) (C5)	Cost (INR) Per kg (C6)	Supply Lead Time (In week) (C7)
Surface Hardened Alloy Steel (M6)	SAE 8620	542	229	1160	680	1850	60	6
Carburised steel (M7)	SAE 8620	647	297	1500	920	2300	60	5
Nitrided steel (M8)	EN40B	693	297	1250	760	1250	72	5

Table:01

5. Determination of Fatigue life of Three Best Gear Materials Considering Technical Aspect:

5.1 CALCULATING THE FATIGUE LIFE OF THREE BEST MATERIALS:
- Gear teeth act as a cantilever beam and it undergoes a fluctuating load so the bending fatigue limit is considerable.
- Manufacturing of gears undergoes a heat treatment process which is case hardening so surface fatigue limit of those materials are not to be considerable.

- For ferrous materials like steel, S-N curve becomes asymptotic at 10^6 cycles, which indicates the stress amplitude corresponding to infinite number of stress cycles. The magnitude of this stress amplitude at 10^6 cycles represents the endurance limit of the materials.
- Considerable load of this experiment is 1000 MPa.

❖ Carburised Steel (M7)

CONSTRUCTION OF S-N DIAGRAM

- ✓ σ_{ut} (MPA) = 2300
- ✓ $0.9\ \sigma_{ut}$ = 2070
- ✓ $Log10(0.9\ \sigma_{ut})$ = 3.31
- ✓ σ_e = 920
- ✓ $Log10(\sigma_e)$ = 2.96
- ✓ Load = 1000 MPa

CALCULATION NUMBER OF CYCLE

$$EF = \frac{DB \times AE}{AD} = \frac{(6-3) \times (3.31-3)}{(3.31-2.96)}$$

$$= \frac{0.93}{0.35} = 2.66$$

$Log_{10}N = 3 + EF$

$= 3 + 2.66 = 5.66$

N = 457008 CYCLE

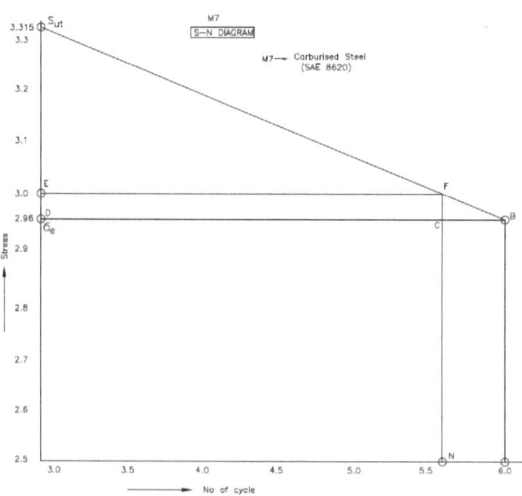

Fig:01

- ### ❖ Nitrided Steel (M8)

CONSTRUCTION OF S-N DIAGRAM

- ✓ σ_{ut} (MPA)   =1250
- ✓ $0.9\, \sigma_{ut}$ =1125
- ✓ $\text{Log}10(0.9\, \sigma_{ut})$ =3.05
- ✓ σ_e =760
- ✓ $\text{Log}10(\sigma_e)$ =2.88
- ✓ LOAD   =1000 Mpa

CALCULATION NUMBER OF CYCLE

$$EF = \frac{DB \times AE}{AD} = \frac{(6-3) \times (3.05-3)}{(3.05-2.88)}$$

$$= \frac{0.15}{0.17} = 0.88$$

$\text{Log}_{10}N = 3 + EF$
$= 3 + 0.88 = 3.88$
N=7585 CYCLE

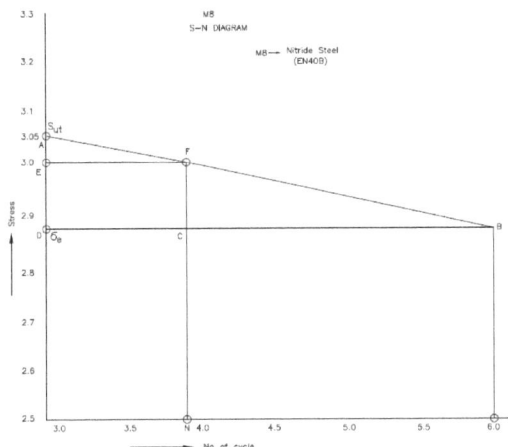

Fig:02

- ❖ **Surface Hardened Alloy Steel (M6)**

CONSTRUCTION OF S-N DIAGRAM

- ✓ σ_{ut} (MPA) = 1850
- ✓ $0.9\ \sigma_{ut}$ = 1665
- ✓ $Log10(0.9\ \sigma_{ut})$ = 3.22
- ✓ σ_e = 680
- ✓ $Log10(\sigma_e)$ = 2.83
- ✓ LOAD = 1000 MPa

CALCULATION NUMBER OF CYCLE

$$EF = \frac{DB \times AE}{AD} = \frac{(6-3) \times (3.22-3)}{(3.22-2.83)} = \frac{0.66}{0.39} = 1.69$$

$Log_{10}N = 3 + EF$
$= 3 + 1.69 = 4.69$
N = 489775 CYCLE

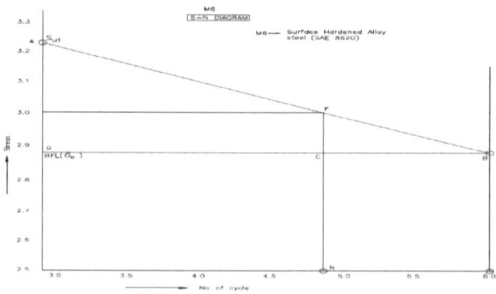

Fig:03

5.2 COMPARATIVE ANALYSIS ON NUMBER OF CYCLE OF GEAR MATERIALS USING S-N DIAGRAM:

MATERIAL	RANK	NO. OF CYCLE
Carburised Steel (M7)	1	457008 CYCLE
Nitrided Steel (M8)	2	7585 CYCLE
Surface Hardened Alloy Steel (M6)	3	489775 CYCLE

Table-4

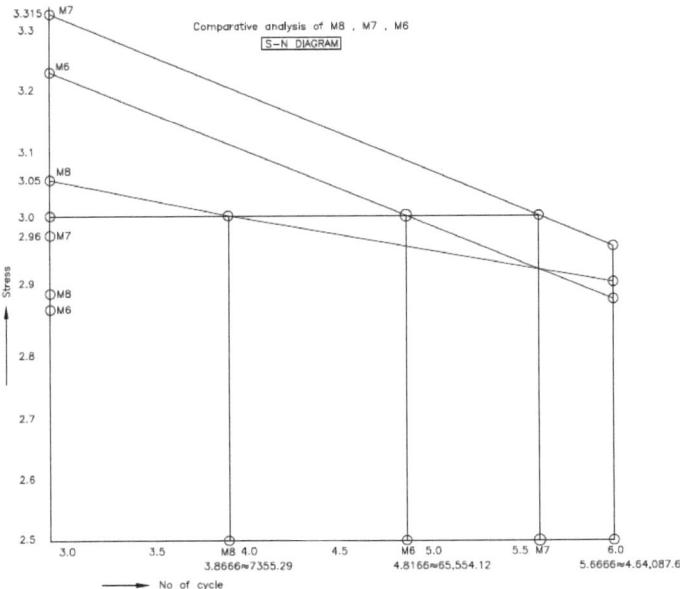

Fig:04

6. SELECTION OF BEST MATERIAL

6.1 FORMATION OF DEVLOPED MATRIX:

MATERIAL	Grade	Surface (Bhn) Hardness (C1)	Core (Bhn) Hardness (C2)	Surface Fatigue Limit (MPa) (C3)	Bending Fatigue Limit (MPa) (C4)	UTS (MPa) (C5)	Cost (INR) Per kg (C6)	Supply Lead Time (In week) (C7)	Number of cycles (C8)
Surface Hardened Alloy Steel (M6)	SAE 8620	542	229	1160	680	1850	60	6	489775 CYCLE
Carburised steel (M7)	SAE 8620	647	297	1500	920	2300	60	5	457008 CYCLE
Nitrided steel (M8)	EN40B	693	297	1250	760	1250	72	5	7585 CYCL

Table-02

TOPSIS METHOD BY USING MATLAB:

RESULT:

The weighted values are:

   0.1059   0.1080   0.1089   0.1100   0.1149   0.1022   0.1001   0.2501

The weighted values got from entropy method

STEP1: Determination of normalized decision matrix

0.7821 0.7710 0.7733 0.7391 0.8043 0.8333 0.8333 0.1122

0.9336 1.0000 1.0000 1.0000 1.0000 0.8333 1.0000 0.0158

1.0000 1.0000 0.8333 0.8261 0.5435 1.0000 1.0000 1.0000

STEP 2:

Determination of positive ideal solution: taking the maximum values of each column from the normalized decision matrix

1 1 1 1 1 1 1 1

Determination of negetive ideal solution: taking the minimum values of each column from the normalized decision matrix

0.7821 0.7710 0.7733 0.7391 0.5435 0.8333 0.8333 0.0158

STEP 3:

Calculation of the separation measure from the positive ideal solution(d_{i_Plus})

0.4805

0.4955

0.1740

Calculation of the separation measure from the negetive ideal solution(d_{i_Minus})

0.1007

0.2188

0.5096

STEP 3: Calculation of R_i

0.1732 0.3064 0.7454

Arranging the final value in descending order:--------->>>

M8 > M7 > M6

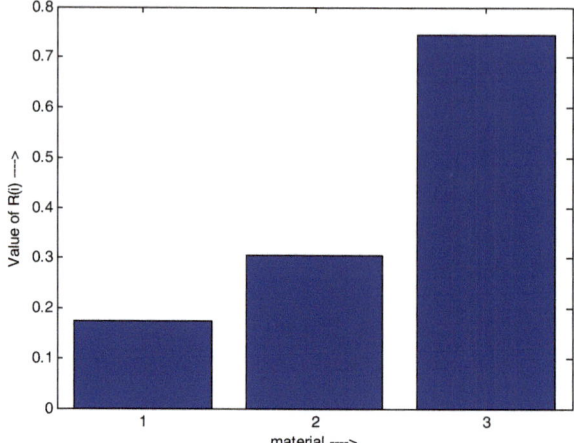

Fig: 05

7. CONCLUSION

It is quite clear that selection of a proper Gear Materials for a given manufacturing application involves a large number of considerations. The use of TOPSIS method is observed to be quite capable and computationally easy to evaluate and select the proper material from a given set of alternatives. These methods use the measures of the considered criteria with their relative importance in order to arrive at the final ranking of the alternative Gear Materials. Thus, these popular MCDM methods can be successfully employed for solving any type of decision-making problems having any number of criteria and alternatives in the manufacturing domain. Use of MATLAB software makes MCDM problem simple and gives prompt results which is very essential in today's decision making environment.

As far as design is concern fatigue life is very much important factor that influence the overall working life of the machine as well as the performance efficiency throughout its life span.

2.5 Edge detection using fuzzy logic

Introduction

- An edge in an image is defined a boundary or contour where an abrupt change occurs in some physical aspect such as gray level value of an image.
- Edge checking is one of the most important tasks in image processing and high level processing.
- Different type of edge operators such as Sobel, Robert, Prewitt are used, this operator depends on the gradient value.

EDGE DETECTORS

- An edge is a vector variable with two components, magnitude and direction. The edge magnitude is the magnitude of the gradient, and edge direction φ is rotated with respect to the gradient direction ψ by – 90 degrees.
- The gradient magnitude and gradient direction are continuous image function calculated as:

$$|grad\ g(x,y)| = \sqrt{\left(\frac{\partial g}{\partial x}\right)^2 + \left(\frac{\partial g}{\partial y}\right)^2}$$

$$\psi = \arg\left(\frac{\partial g}{\partial x}, \frac{\partial g}{\partial y}\right)$$

CLASSIFICATION OF EDGES

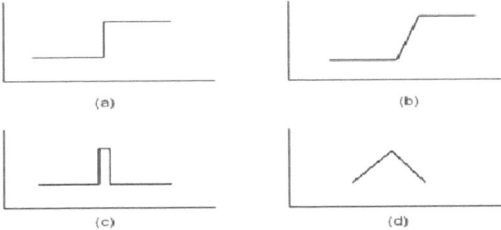

Fig: (a) step edge (b) ramp edge (c) line edge (d) roof edge

STEPS IN EDGE CHECKING

- ❏ The overviews of the steps in edge checking are as follows:

  ➢ **Filtering**:- Images are often corrupted by random variation in intensity values, called noise. To reduce the noise, filtering is done. However, there is trade-off between edge strength and noise reduction.

Enhancement:- in order to facilitate the checking of edges, it is essential to determine changes in intensity in neighborhood of a point. Enhancement emphasizes pixels where there is a significant changes in local intensity value

  ➢ **Detection/checking**:- Many points in an image have a nonzero value for the gradient, and not all of these points are edges for a particular application. Some method should be used to determine which points are edge points. Frequently, thresholding provides the criterion used for detection.

CLASSIFICATION OF EDGE CHECKER

- ❏ Operators are able to detect edge direction are represented by a collection of masks, each corresponding to a certain direction. Generally three types operators are used in edge checking.

  ➢ Robert operator.
  ➢ Sobel operator
  ➢ Prewitt operator

Robert operator

The Robert operator performs a simple, quick to compute, 2-D spatial gradient measurement on an image

+1	0
0	-1

0	+1
-1	0

Figure:Robert Mask

- The primary advantage of the Robert operators is high sensitivity towards noise, because very few pixels are used to approximate gradient.

Original image

Using Robert operator

Prewitt operator

- The Prewitt edge detector is an appropriate way to estimate the magnitude and orientation of an edge. The prewitt operator is limited to 8 possible orientations. Convolution mask of Prewitt operator.

-1	+1	+1
-1	-2	+1
-1	+1	+1

0 deg

+1	+1	+1
-1	-2	+1
-1	-1	+1

45 deg

Fig : Prewitt mask

- The direction of the gradient is given by the mask giving maximal response. This is also the case for all the following operators approximating the first derivative.

Original image

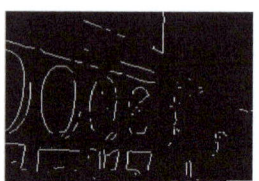
Using Prewitt operator

Sobel operator

> The sobel operator performs a 2-D spatial gradient measurement on an image and so emphasizes region of high spatial frequency that correspond to edges. The convolution mask of the Sobel operator are given below

-1	0	+1
-2	0	+2
-1	0	+1

+1	+2	+1
0	0	0
-1	-2	-1

Gy
Gx

Fig: sobel Mask

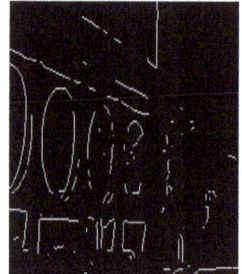

Original image   Using sobel operator

Cause of use fuzzy logic:

- ❑ Edge checking operator such as Robert, Prewitt, Sobel are the classical techniques have some drawbacks:
- ➢ some of the classical techniques are used with certain parameters such as threshold and σ to implement edge detection process.
- ➢ The another restrict in classical approach, results generally have fixed edge thickness.

to overcome this sort of problem fuzzy logic is introduced in edge detection.

Fuzzy logic in edge checking

- ➢ Fuzzy approaches for image segmentation can be categorized into four classes:
1. Segmentation via thresholding
2. Segmentation via clustering
3. Supervised segmentation
4. Rule based segmentation.
5. Fuzzy based rules method in most of fuzzy based detection algorithms are used.
6. In these methods, adjacent pixels are assumed in some classes.
7. Fuzzy system inference are implemented using appropriate membership function, defined for each class.

Primary edge checking and fuzzy rules

- ➢ Two different methods, gradient and standard deviation of pixels value.
- ➢ Edges are extracted.
- ➢ Applying fuzzy logic , final decision about whether each pixel is edge or not.
- ➢ The gradient values are computed by Sobel operator and the pixels with gradient value bigger than a threshold are edge candidate.

- Similarly pixels with standard deviation (SD) greater than a threshold value are edge candidate. Each pixel SD is computed by 3*3 mask.

P_1	P_2	P_3
P_4	P_5	P_6
P_7	P_8	P_9

- ✘ Two computed values are used as fuzzy inputs. The final decision about the edge candidate pixels is based on the output of fuzzy system.
- ✘ Appropriate membership functions are defined for fuzzy system inputs.
- ✘ To apply this function, first both SD and gradient values are mapped to the range [0 100].

Membership function

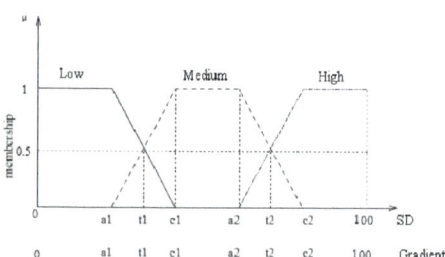

[0 c1]= SD_L

[a_1 c_2] = SD_M

[a_2 100] = SD_H

- The output of fuzzy system explains to how extent a pixel could be edge.
- By the fuzzy rules, the output of this system is classified to one of three classes.
- The first class, E_L, correspond to pixels with low probability value to belong to edge pixels set.
- E_M corresponds to medium probability

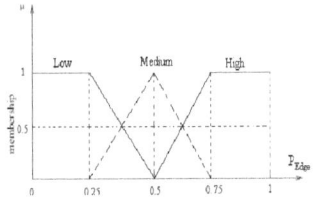

Output membership function

0 to 0.5- low prob. Pixel.

0.25 to 0.75- medium prob. Pixel.

0.5 to 1- high prob. Pixel

algorithms

- If SD value of a pixel being equal to K_1, and gradient be equal to K_2, the fuzzy rules are defined as following:

1. If K_1 in SD_L and K_2 in G_L then P_{edge} classified to E_L.
2. If K_1 in SD_L and K_2 in G_M then P_{edge} classified to E_L.
3. If K_1 in SD_L and K_2 in G_H then P_{edge} classified to E_M.
4. If K_1 in SD_M and K_2 in G_L then P_{edge} classified to E_L.
5. If K_1 in SD_M and K_2 in G_M then P_{edge} classified to E_M.
6. If K_1 in SD_M and K_2 in G_M then P_{edge} classified to E_H.
7. If K_1 in SD_H and K_2 in G_L then P_{edge} classified to E_M.
8. If K_1 in SD_H and K_2 in G_L then P_{edge} classified to E_H.
9. If K_1 in SD_H and K_2 in G_H then P_{edge} classified to E_H

- **The 9 rules are summarized as:**

Grad SD	Low	Medium	High
Low	E_L	E_L	E_M
Medium	E_L	E_M	E_H
High	E_M	E_H	E_H

Defuzzification

✘ Defuzzification is done using the following equation:

Where $P_{edge}(j)$ is the pixel membership value in the jth class, and C_j is the jth output class center

Original image

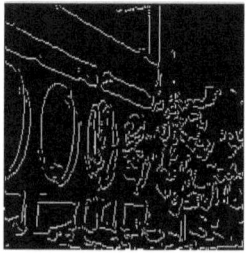

Applying fuzzy rules

Result comparison

➢ More prominent edges.

➢ Detect more no. edges.

➢ More information

Industrial application of edge detection

- Product assembly & inspection, pattern recognition.
- In machine vision gauging.
- Aircraft & automobile industries.
- In security industry.
- In glass & plastic industries for accurate measurement.

conclusion

- Based on primary edge checking methods, gradient and SD computed each pixel, and are used as fuzzy input, define appropriate membership function.
- Result show high quality and superiority of the extracted edges compared to other classical methods

4.8 Design of Inspection Strategy –an Example

In industry, mainly in oil refinery & petrochemical sectors are looking over a problem for improper DFT(Dry Film Thickness) in large number of area (piping & tank maintained) painting system by third party. Result, consumer's risk is gradually increased.
In this research venture, developed a procedure to draw an O.C curve using Minitab software to minimize the consumer's risk on external & internal in a large area of tank & pipe line painting to resist corrosion by best expert's review using MCDM method.
The paper deals with large area coating system by measuring NDT methods (DFT) to improve the quality level as well as better inspection strategy.

Statement of the problem
Basically, in oil refinery/petrochemical based industry required large number of area(tank and pipe line) to come under painting system to resists corrosion. This type of project done by third party. To maintain the quality control by DFT checking responsibilities are vested on the department of inspection. Practically, this is not possible to 100% inspection. An inspector inspected this type of jobs by his experience and applies Standards Used for DFT Measurement. Lack of 100% inspection, improper coating found many places, Result, gradually increase the risk to failure the coating system.

Acceptance sampling is an inspection procedure used to determine whether to accept or reject a specific quantity of material. It is a part of operation management or of accounting, auditing and services quality supervision. Acceptance sampling is most likely to useful in the following situation:
1) When the time and cost of 100% inspection is extremely high.

2) When 100% inspection is not technologically feasible or would required.
3) When there are many items or spots to be inspected and the inspection error rate is sufficiently higher than 100% inspection might cause a higher percentage of defectives (here, low DFT in painting inspection) to be passed than would occur with the use of sampling plan.

Dry film thickness (DFT) is probably the single most important measurement made during inspection or quality control of protective coating application. Even the most basic protective coating specification will inevitably require the DFT to be measured. It is considered to be the most important factor determining the durability of a coating system. The thickness of each coating layer in a system and the total system DFT will have to be measured and recorded to show that the specified system will meet the desired durability.
In industry, Standards Used for DFT Measurement: SSPC-PA 2, AS 3894.3, ISO 19840, PSPC.

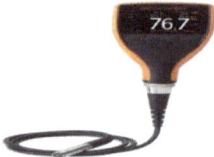

Figure-1:Elcometer:DFT measurement machine

Time study is a direct and continuous observation of a task, using a timekeeping device (e.g., decimal minute stopwatch, computer-assisted electronic stopwatch, and checkingvideotape camera) to record the time taken to accomplish a task[3] and it is often used when:
- there are repetitive work cycles of short to long duration,
- wide variety of dissimilar work is performed, or process control elements constitute a part of the cycle.

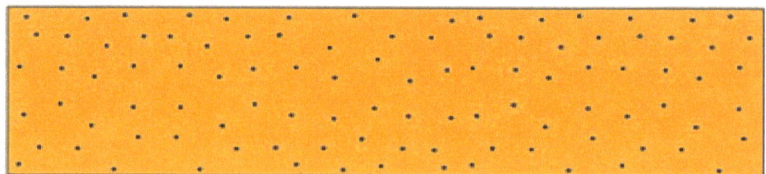

Figure -2: coating thickness measurement spot on flat area of tank plate.

Multiple criteria decision making (MCDM) is the process of selecting the best alternative from a set of feasible alternatives considering multiple conflicting criteria. In precise terms criteria are considered to be 'strictly' conflicting if the increase in satisfaction of one results in a decrease in satisfaction of the other. An MCDM process always contains at least two alternatives and two conflicting criteria (Bhattacharya et al., 2003). MCDM are divided two broad categories:

Multiple Attribute Decision Making (MADM) and Multiple Objective Decision Making (MODM).

Research Design Framework

Figure -3

2. **Mathematical Model:**

PART-1
Taken the 732 square meters area painting inspection, we, fixed the inspection spot in 100spot/sq.meter and get a lot size 73200 units

- Existing Sampling plan adapted by various experienced expert in IOCL to draw a O.C curve in this case study as given below:

SL. NO.	LOT SIZE (N)	EXPERTS	SAMPLE SIZE (n)	ACCEPTANCE NUMBER (C)
1	73,200	A	1600	50
2	73,200	B	1200	20
3	73,200	C	2000	20
4	73,200	D	1000	10

Table-1:

- For calculation of probabilities of acceptance, a Poisson's Distribution method is used as given below.
  Drawing an O.C curve on existing sampling plan:
  Lot size (N) =73,200
  Sample size (n) =1600,1200,2000,1000 etc.by various experts.

- Inspected the sample of lot size and if number of defectives (here, low DFT) are equal to Acceptance number(C) then accept the DFT or if it is more than C (i.e. =1, 2, 3...) then it will reject the lot.

We have to draw an O.C curve for this assume the
Percent of defectives in a lot as, P' (% defectives) =0.1%, 0.5%, 0.8%, 1.0%, 2.0%, 4.0%, 6.0%, 10.0%

According to Expert "A"

N=73200, n=1600,C=50
We consider, simplifying the calculation all data divided by 100 and we get
N=732,n=16,C=0

SL NO.	Percent of defectives (P')	np'	Probability of Acceptance (P_a)
1	0.1%	0.02	0.98
2	0.5%	0.08	0.92
3	0.8%	0.12	0.88
4	1.0%	0.16	0.85
5	2.0%	0.32	0.72
6	4.0%	0.64	0.52
7	6.0%	0.96	0.38
8	10.0%	1.60	0.20

Table-2: Calculation of single sampling plan for Expert-A

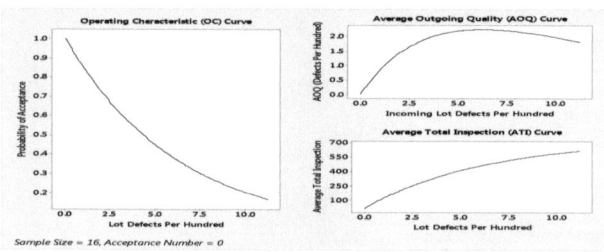

Figure-5: O.C. curve for Expert-A

If we consider lot tolerance percent defectives as 5% corresponding consumer's risk can be calculated.
From fig-1, if we consider LTPD= 5%. The Consumers risk is 44.9%

Hence consumer's risk is 44.9% with the existing sampling plan. It must be minimized.

Similarly,
According to Expert -B

N=732,n=12,C=0

SL NO	P'	np'	P_a
1	0.1%	0.01	0.99
2	0.5%	0.06	0.94
3	0.8%	0.10	0.91
4	1.0%	0.12	0.89
5	2.0%	0.24	0.77
6	4.0%	0.48	0.61
7	6.0%	0.72	0.48
8	10.0%	1.20	0.30

Table-3: Calculation of single sampling plan for Expert-B

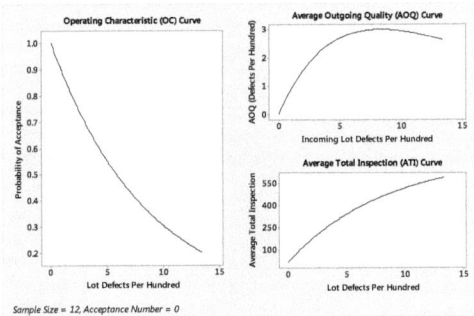

Figure-6: O.C. curve for Expert-B

From fig if we consider LTPD= 5%. Consumers risk is 54.9%
Hence consumer's risk is 54.9% with the existing sampling plan. It must be minimized.

According to Expert -C
N=732,n=20,C=0

SL NO.	P'	np'	P_a
1	0.1%	0.02	0.98
2	0.5%	0.10	0.90
3	0.8%	0.16	0.85
4	1.0%	0.20	0.82
5	2.0%	0.40	0.67
6	4.0%	0.80	0.45
7	6.0%	1.20	0.30
8	10.0%	2.00	0.13

Table-4: Calculation of single sampling plan for Expert C

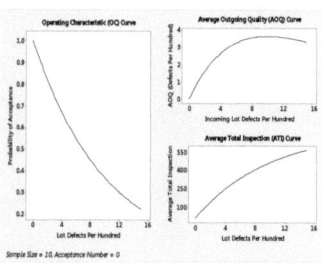

Figure-7: O.C. curve for Expert-C

From fig if we consider LTPD= 5%. Consumers risk is 36.8%Hence consumer's risk is 36.8%with the existing sampling plan. It must be minimized.

According to Expert "D"
N=732,n=10,C=0

SL NO.	P'	np'	P_a
1	0.1%	0.01	0.99
2	0.5%	0.05	0.93
3	0.8%	0.08	0.92
4	1.0%	0.10	0.90
5	2.0%	0.20	0.81
6	4.0%	0.40	0.67
7	6.0%	1.60	0.55
8	10.0%	1.00	0.368

Table-5: Calculation of single sampling plan for Expert -D

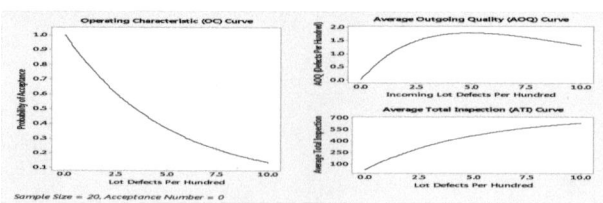

Figure-8: O.C. curve for Expert -D

From fig if we consider LTPD= 5%. Consumers risk is 60.7%
Hence consumer's risk is 60.7%with the existing sampling plan. It must be minimized.

Limitations and problems in existing sampling plan:
1. Lot tolerance percent defectives are larger.
2. As this sampling plan is based upon single sampling, there may be possibility of acceptance of defective lots.
3. As acceptance number C=0, vendors may be supply by considering this acceptance number. However
some tolerance limit should be given to the vendors for better coordination and commitment.
4. Consumer's risk in a single sampling plan is always larger and it is acceptance of such a lot, which would have been rejected. It can affect the consumer and his next production work and assembly work.

Possible Remedies for Overcoming the Limitations:

1. If increases the sample size and increases the acceptance number lot tolerance % defectives can be
Minimized. But still it is single sampling hence there may be possibility of acceptance of defective components.
2. By using single sampling inspection of various experts, comparatively analysis to determine the best minimized LTPD of this expert's decision.
3. Using the another optimization method MCDM to calculate the ranking the best expert in respect of sample size, consumer risk, and working time to minimized as well as optimized the inspection system.

PART-2
7.4.1. Simple additive weighting (SAW)
Step 1 Formation of decision matrix: Criterion outcomes of decision alternatives can be collected in a table called Decision Matrix comprised of a set of columns and rows. The matrix rows represent decision alternatives, with matrix columns representing criteria. A value found at the intersection of row and column in the matrix represents a criterion outcome - a measured or predicted performance of a decision alternative on a criterion. The decision matrix is a central structure of the MCDA/MCDM since it contains the data for comparison of decision alternatives.

$$X = \begin{array}{c} \\ A_1 \\ \vdots \\ A_i \\ \vdots \\ A_m \end{array} \begin{array}{c} C1 \quad\quad CJ \quad\quad Cn \\ \begin{bmatrix} x_{11} & \cdots & x_{1j} & \cdots & x_{1n} \\ \vdots & \cdots & \vdots & \cdots & \vdots \\ x_{i1} & \cdots & x_{ij} & \cdots & x_{in} \\ \vdots & \cdots & \vdots & \cdots & \vdots \\ x_{m1} & \cdots & x_{mj} & \cdots & x_{mn} \end{bmatrix} \end{array}$$

.........(1)

x_{ij} is the performance rating of alternative i with respect to criterion j,
A_j is ith alternative, C_j is the jth criterion

Step 2 Formation of Weight Matrix:

Different importance weights to various criteria may be awarded by the decision makers. These importance weights forms the weight as follows.

$$W = [W_1 \cdots W_j \cdots W_n] \quad \ldots\ldots (2)$$

Step 3 Normalization of performance rating
Units and dimensions of performance ratings of columns under criteria differ. For the purpose of comparison, these performance ratings are converted into dimensionless units by normalization using following equations:

$$\bar{x}_{ij} = \frac{x_{ij}}{\max_i(x_{ij})} \quad \text{for benefit criteria} \ldots\ldots(3)$$

$$\bar{x}_{ij} = \frac{\min_i(x_{ij})}{x_{ij}} \quad \text{for non-benefit criteria}\ldots(4)$$

Normalized decision matrix

Step 4: Composite score:
Computation of composite score (CSi) for alternatives.

$$CS_i = \sum_{j=1}^{n} (\bar{w}_j * \bar{x}_{ij})$$

Step 5 Ranking and selection of best alternative:

Ranking of products in descending order of composite scores (CSi)

$$\bar{X} = \begin{matrix} A_1 \\ A_2 \\ \vdots \\ A_m \end{matrix} \begin{bmatrix} \bar{x}_{11} \cdots & \ldots \bar{x}_{1j} \cdots & \bar{x}_{1n} \\ \bar{x}_{i1} \cdots & \ldots \bar{x}_{ij} \cdots & \bar{x}_{in} \\ \vdots & \vdots & \vdots \\ \bar{x}_{m1} & \bar{x}_{mj} & \bar{x}_{mn} \end{bmatrix}_{m \times n} \quad \ldots\ldots\ldots (5)$$

➤ **. Formation of Matrix comparing with other Inspection Factors:**

SL. NO.	SAMPLE SIZE(n)	CONSUMER RISK (L.T.P.D) [5%]	TIME (TIME STUDY METHOD)
EXPART-A	16	44.9%	7min.

EXPART-B	12	54.9%	5.4min.
EXPART-C	20	36.8%	9min.
EXPERT-D	10	60.7%	4.5min.

Table-6

> **Computational Result by Mat lab software:**

The weighted values are:

Criteria-1	Criteria-2	Criteria-3
0.3381	0.3222	0.3397

Table-7: weighted values by Entropy method

✓ **Result of the SAW method:**

The values of (s) are:

A	B	C	D
0.7529	0.7019	0.8301	0.7041

Table-8: Final result in ranking

Arranging the final value in descending order: E3>E1>E4>E2

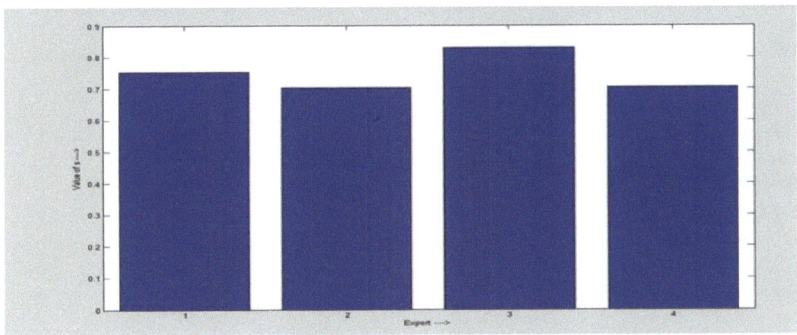

Figure-9: Ranking best Expert by MCDM.

Conclusions:
The design of acceptance sampling process of this particular job includes decisions about sampling versus complete inspection. In this design strategy has shown prominently the quality

risk through the graphs individually for each inspection experts, using Minitab software, that can easily and more quickly calculate and draw operating characteristic curve.

The second level optimization tool, MCDM is used for selection the best expert comparing with other factors using Mat lab software that can easily shown the graph for best decision through expert's ranking.

This model helps the large number of painting inspection become automated as well as less time consuming.

4.9 IE approach in NDT environment-An Example

In industrial storage tank, specifies minimum acceptable thicknesses for the bottom & annular plate [Figure: 1] by API 653. It indicates that the remaining thicknesses may be quantified using either probabilistic or deterministic method. The bottom plate thicknesses are then compared to the required thicknesses to determine their acceptability.

The deterministic method uses more extensive inspection data to quantify the remaining thickness of the bottom. The required data include the following:

- Original plate thickness.
- Average and maximum depth of internal and underside pitting.
- Maximum internal and underside pitting rates and maximum general corrosion rate.
- Average depth of general corrosion.
- The time until the next inspection.

But in practical approaches of oil & petrochemical industry, there are various manual inspection problems occur for the hazardous condition like Tank, Furness, boiler etc.

Often it is observed that the report was not correct. For better decisions, introducing the inspection accuracy on alternative/cross check policy should be implemented by execution to analysis the inspection accuracy.

The model uses the fraction defect rate in the inspection batch, the cost of inspection per item which is inspected, and the cost of damage that one defective plate would cause if it were not inspected properly. The total cost per plate 100% inspection can be formulated.

Figure: 1

1.1 Inspection accuracy: Refers to the capability of inspection process to avoid these types of errors; Measures of inspection accuracy are suggested by DRURY for the case in which parts are classified by an inspector. Inspected items of good quality are incorrectly classified as not conforming to accepted specification, and nonconforming items are mistakenly classified as conforming. These two kinds of errors are called "False alarm" and "Miss" respectively.

Two types of errors in cross check inspection:

	Conforming Item	Nonconforming Item
Accepted Item	Right decision	Miss
Rejected Item	False alarm	Right decision

1.2 Inspection or No inspection: A model for deciding to inspect at a certain point in the production sequence is proposed in Juran and Gryna. The model uses the fraction defect rate in the inspection batch, the inspection cost per unit inspected, and the cost of damage that one defective unit would cause if it were not inspected.

2. STATEMENT OF THE PROBLEM

During tank maintenances, the bottom plate metal loss checking is an important part of tank inspection. Sometimes error occurs in the inspection procedure such as those plates of good quality are incorrectly classified and vice-versa. In manual inspection, these errors result from factors such as:
  i. Inherent variations in the inspection procedure
  ii. Complexity, hazard and other various difficulty of the inspection task
  iii. Inaccuracy in measuring instrument
  iv. Mental fatigue etc.

For this irresolute inspection errors, It's have possible chance to leak [Figure: 02 & 03] of the tank before the next tank Maintenance & Inspection (M & I), results-

- Production loss
- Drain of money
- Idle times are increase etc.

Figure: 2 Failure sample bottom plate   **Figure: 3** Failure sample annular plate

On this critical inspection cases the inspection accuracy is very important. But these types of situation the exact accuracy as well as defect rate remain undetermined. Whenever, the leak is found for failure of inspection then it is impossible to assess the past inspection accuracy on the basis of inspected data. Modeling a system to make automated inspection accuracy for minimizing the decision criticality. For minimizing this type of errors, implemented the alternative/cross check inspection are frequently used [Figure: 4].

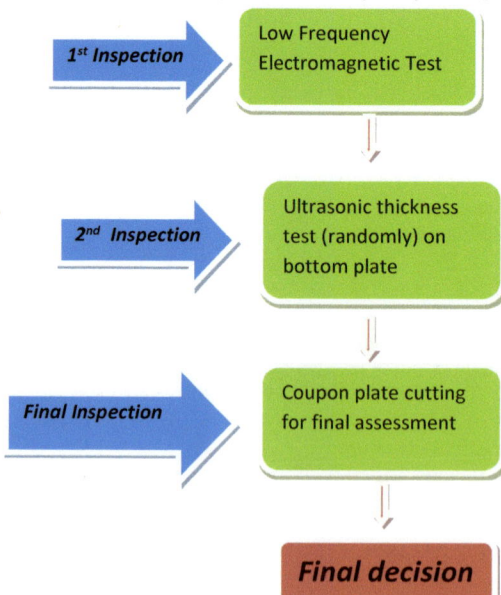

3. RESEARCH OBJECTIVES

In practical approaches, during LFET on bottom floor of a tank inspection, frequently errors are come out for hazarded condition. To avoid this kind of situation a different approach has been implemented i.e. coupon plate cutting for sample inspection, Thickness checking on spot sampling etc. As a result, processing time of inspection is maximized, risk factor is unpredictable, No future inspection can be done because of increasing the inspection cost etc.

In this research, addresses to analysis the inspection accuracy by

I. Automated computational methodology to determine the accuracy in cross check approaches.
II. Finding the errors for future inspection
III. Simplifying the decision about the plates will be replaced or not.

4. MODELING INSPECTION SYSTEM FLOW DIAGRAM

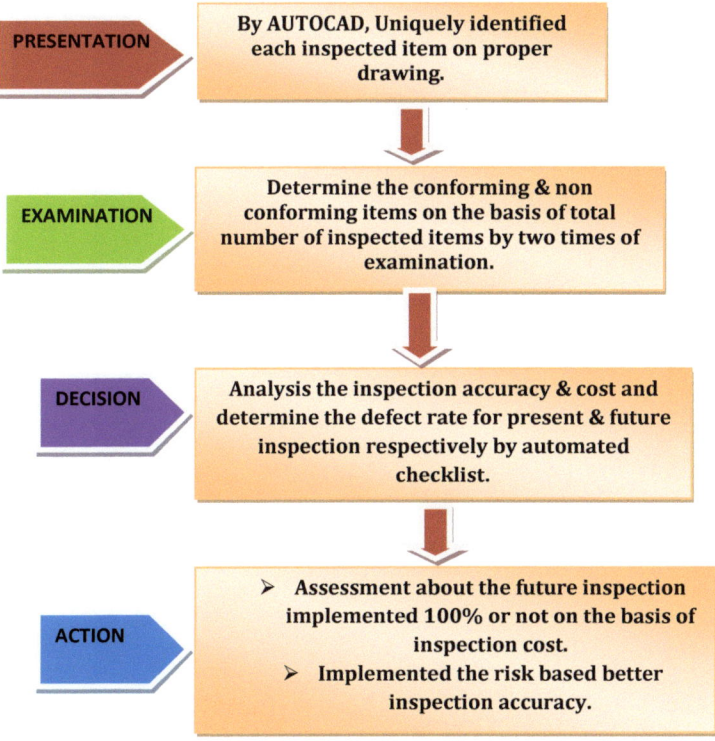

5. EXPERIMENT

Basically in oil & petrochemical industry looking over an inspection problem in hazarded area like Tank, Furnace, boiler etc. During manual inspection, various types of errors are occurring in inspection time and the inspection report become incorrect by human error. To make the inspection accurate the cross check policy should be implemented by execution.

During storage tank maintenance, the bottom plate thickness test is a part of critical inspection, because various types of hazard are occurred that time. Analysis of the inspection accuracy on bottom plate [Figure: 05] thickness during tank M&I under cross check methodology:

5.1 Design & Identify the each inspected plates by numbering through the CAD on complete bottom area.

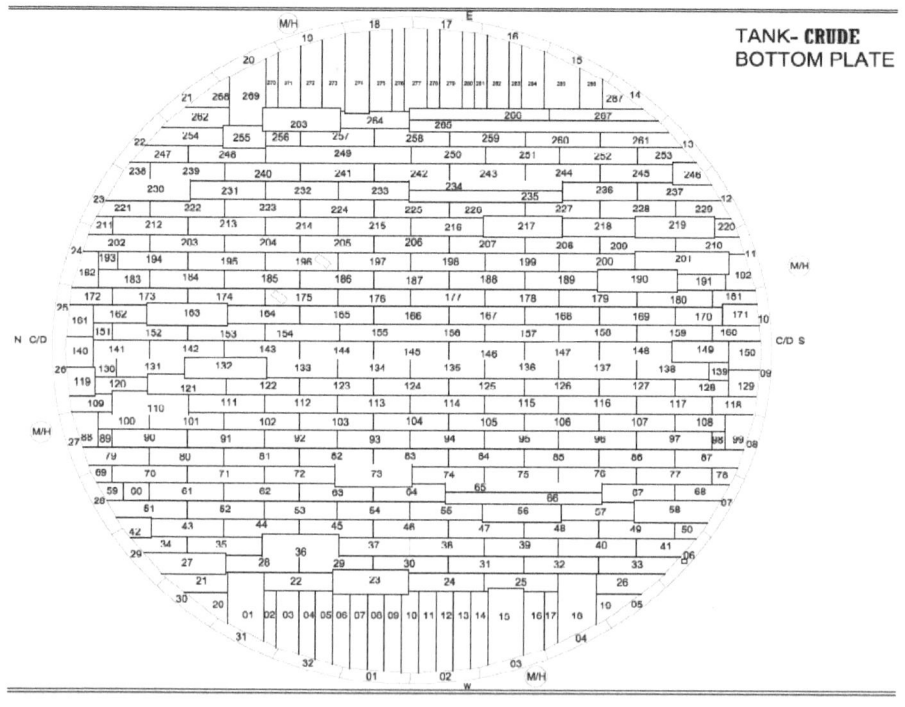

Figure: 05

Using the following drawing:

- Total bottom plates of tank = 287
- Total annular plates of tank = 32

So, Total number of inspected plates of tank = 319

- ✓ Each plate inspection cost (C_s) = INR 950 [approx]
- ✓ Each plate damage cost (C_d) = INR 38,000 [approx]

5.2 Inspection and Data Analysis

From inspection data evaluation standpoint, it would be ideal to have a complete thickness map of the bottom. However, it would be expensive to perform the ultrasonic and LFET inspections that are necessary to do that, and for such an extensive survey of "MAN-MACHINE" error for analysis the inspection accuracy is necessary.

5.2.1 LFET testing is useful and the most reliable technique on rough surfaces or surfaces with wet films where the plates or the pipelines coated. It is used to detect material loss, which caused by corrosion or other deterioration process. The LFET operated scanner moves over the entire surface of the tank, while generating an electromagnetic field into the steel plate. The electromagnetic field generates eddy current in the conductive material. The system measures the changes in the electromagnetic field caused by the generated eddy current. The defects and the corrosion maps are calculated from these collected values. Since the scanner head does not have

❖ In First Inspection (Low Frequency Electromagnetic Test) Report,
  The number of defective plates are detected = 56
  The percentages of good or defective items are shown in following chart [Figure: 06]

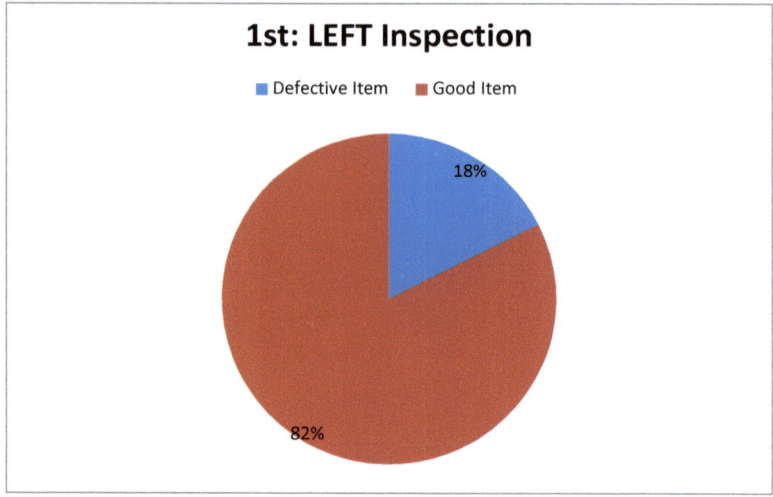

Figure: 06

In hazarded or critical based inspection condition, the second inspection is implemented to minimize the risk.

5.2.2 Ultrasonic nondestructive testing (NDT) – a method of characterizing material thickness, integrity, or other physical properties by means of high frequency sound waves is a widely used technique for product testing [Figure: 8] and quality control. In thickness gauging applications, ultrasonic techniques permit quick and reliable measurement of thickness without requiring access to both sides of a part. Calibrated accuracies as high as ±2 micrometers or ±0.0001 inch are achievable in some applications. Most engineering materials can be measured ultrasonically under proper maps to evaluate the thickness data [Figure: 7]

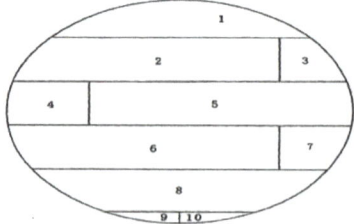

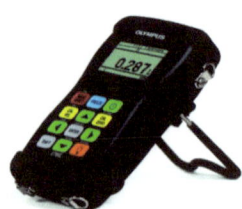

Figure: 7   Figure: 8

- In second Inspection Report, it was found that 16 of these reported defects were in facts good pieces. Whereas a total of 21 defective plates in a tank were undetected through the inspection [Figure: 9]

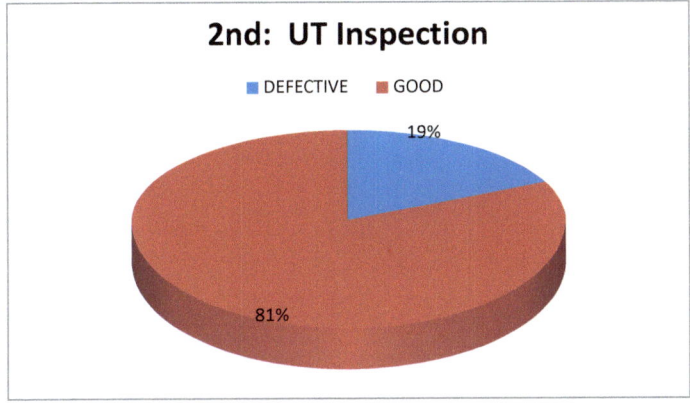

Figure: 9

> Comparative analysis of inspection data in two times different inspection procedure [Figure: 10]

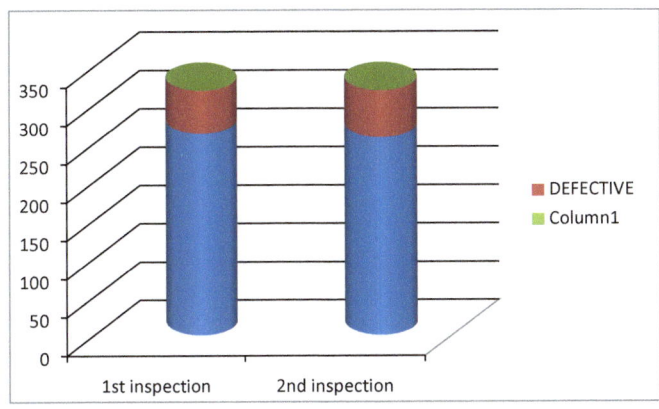

Figure: 10

So, the total "FALSE ALARME" = 16

The total "MISSES" = 21

5.3 Constructing computing logic on the inspection data by Microsoft Excel as following

Microsoft Excel has the basic features of all spreadsheets, using a grid of cells arranged in numbered rows and letter-named columns to organize data manipulations like arithmetic operations. It has a battery of supply functions to answer statistical, engineering and financial needs.

Developed a check sheet by Microsoft Excel [Figure: 11]:

- Actual Acceptance (A) = Good (G) in the 1st inspection + False Alarm (F) during cross check – Miss (M) during cross check.

- Actual Reject (R) = Bad (B) in the 1st inspection + Miss (M) during cross check – False Alarm (F) during cross check

- Probability of conforming item (P_1) = [Total Item (Q) in batch – {Bad (B) in the 1st inspection + Miss (M) during cross check}] / Actual Acceptance (A)

- Probability of Non-conforming item (P_2) = [Total Item (Q) in batch – {Good (G) in the 1st inspection + False Alarm (F) during cross check}] / Actual Reject (R)

- Accuracy = {Probability of conforming item (P_1) + Probability of Non-conforming item (P_2)} / 2

- Defect Rate (q) = (1 - over all inspection accuracy)

- Batch cost for 100% inspection (C_b) = Q.C_S (Q= total parts in a batch) &

  Cs = inspection cost

- Batch cost for NO inspection (C_n) = Q.q. C_d (C_d = inspection damage cost)

- The critical defect value (Q_C) = C_S / C_d [critical value represents the breakeven point between inspection or no inspection]

Total Parts in a Batch	1st inspection		Cross Check		Good Decision		Probability of conforming item	Probability of Non-conforming item	Accuracy	Defect Rate	Inspection Cost	Damage Cost	Batch cost for 100% inspection	Batch cost for NO inspection	the critical defect value
	Good	Bad	False Alarm	Miss	Actual Acceptance	Actual Reject									
Q	G	B	F	M	A	R	P1	P2	A	q	Cs	Cd	Cb	Cb	Qc
100	88	12	4	6	86	14	0.953488372	0.571428571	0.76245847	0.23754153	10	15	1000	356.3122924	0.66666667
20	10	10	6	4	12	8	0.5	0.5	0.5	0.5	6	9	120	90	0.66666667
319	263	56	16	21	258	61	0.937984496	0.655737705	0.7968611	0.2031389	950	38000	303050	2462449.739	0.025

Figure: 11

5.4 RESULT

- The proportion of good parts reported as conforming is (P_1)= 0.938
- The proportion of defective parts reported as nonconforming is (P_2) = 0.656
- The overall inspection accuracy (A) = 0.797 [Figure: 12]
- Defect Rate (q) = 0.203

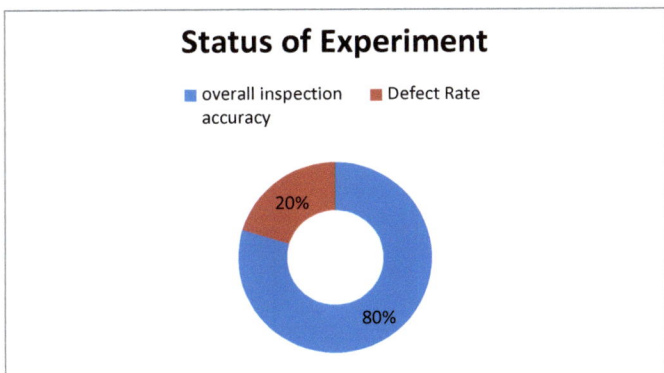

Figure: 12

- In quality assurance approaches, if we consider the overall inspection accuracy is an average accuracy then it show the inspection status as a graph [Figure:03] in Poisson distribution by Minitab software as following:

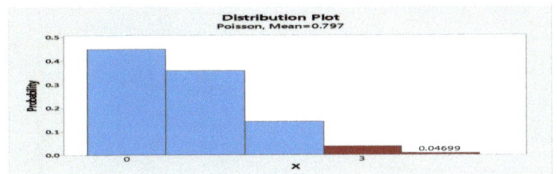

Figure: 03

I. Batch cost for 100% inspection (C_b) = INR 3,03,050
II. Batch cost for NO inspection (C_n) = INR 24,60,766
III. The critical defect value (Q_C) = 0.025

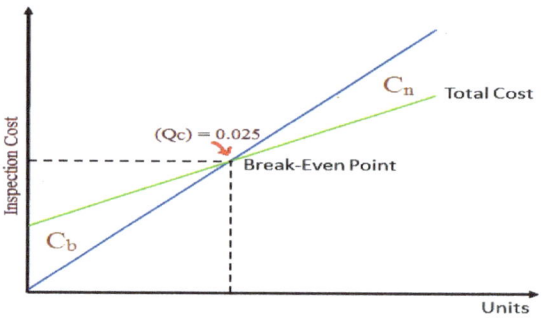

Figure: 04 Break even Analysis

Since the anticipated defect rate in the inspected batch is = 0.203, the decision should be inspect. Observed that this decision is consistent with the two batch costs calculated for no inspection and 100% inspection. The lowest cost is attained when 100% inspection is used.

5.5 DECISION

Based on past history with the inspected items, the batch fraction defect rate q is less than this critical level then no inspection is indicated. On the other hand, if it is expected that the fraction defect rate will be greater than Q_C, then further inspection is necessary.

If, Q_C < q	Inspection is indicated
If, Q_C > q	NO inspection is indicated

In this experiment,

The critical defect value (Q_C) = 0.025 is less than the Defect Rate (q) = 0.203

$$\boxed{Q_C < q}$$

❖ Future or further inspection is indicated

5.6 ADVANTAGES OF THIS METHODOLOGY

- Determine the inspection accuracy easily in excel sheet.
- Assessment of the inspector's responsibility on the particular job
- Analysis the risk based inspection report
- Inspections are less time consuming
- Determine the decision about future inspection under cost based inspection.
- Minimize the human error

CONCLUSION

The inspection methodology is used to determine the error rate for quality control during critical inspection environment. It also helps to evaluate the risk for future inspection by automated data record. The risk based inspection policy is merged on cost based inspection methodology to analysis the next inspection criticality. The complete execution process is made on computationally as well as less time consuming. This model useful for various inspection procedures as well as prediction of human error.